IMAGES
of America

THE SEABEES AT PORT HUENEME

ON THE COVER: Esther Williams, star of MGM's *Bathing Beauty*, sits on the lap of John "Red" Purcell at the official opening of two water training tanks at Camp Rousseau on July 13, 1944. A huge audience of Seabees and military and civilian personnel came out to see Esther Williams and MGM's twin ballet swimming and water waltzing team, Martha and Patsy Brown.

IMAGES
of America

THE SEABEES AT
PORT HUENEME

Gina Nichols

ARCADIA

Published by Arcadia Publishing
Charleston SC, Chicago IL, Portsmouth NH, San Francisco CA

Printed in the United States of America

Library of Congress Catalog Card Number: 2006920412

For all general information contact Arcadia Publishing at:
Telephone 843-853-2070
Fax 843-853-0044
E-mail sales@arcadiapublishing.com
For customer service and orders:
Toll-Free 1-888-313-2665

Visit us on the Internet at www.arcadiapublishing.com

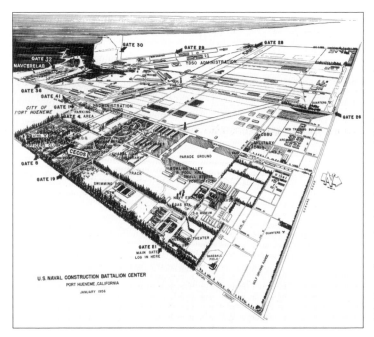

U.S. Naval Construction Battalion Center Port Hueneme is the subject of this 1956 illustration. On March 9, 1942, the Navy officially served notice that it intended to build an advance base depot at Port Hueneme as an emergency naval base. Construction began on March 12, 1942. The base would cover approximately 1,700 acres and serve as a major shipping hub for the Navy during its battles in the Pacific.

CONTENTS

ACKNOWLEDGMENTS

The author wishes to thank the Naval Facilities Engineering Command, the Naval Construction Forces, the CEC/Seabee Archive and the U.S. Navy Seabee Museum, for without their enthusiastic support of Seabee history this book could not have been written. All of the photographs in this book are from the CEC/Seabee Archive and the Naval Media Center's Navy NewsStand. The CEC/Seabee Archive, located at Construction Battalion Center Port Hueneme since its inception in 1962, is the official repository for the U.S. Naval Construction Force records. All of the photographs presented here were taken by official U.S. Navy photographers or are part of personal collections donated to the archive with no restrictions.

Special thanks to my son Hunter, who endured endless peanut butter and jelly sandwiches and innumerable hours at the CEC/Seabee Archive. Also, thanks to my parents and friends who patiently and tirelessly supported me in this venture.

INTRODUCTION

The U.S. Naval Construction Battalion Center at Port Hueneme, California, best known as the "Home of the Pacific Seabees," developed in early 1942 from a need for another major Pacific Coast shipping port. The Navy dispatched survey teams in February 1942, to locate land for an advance base depot on the West Coast from which men, construction materiel, and equipment could be shipped to the Pacific Theater. Port Hueneme, with its deep-water harbor and wide expanse of undeveloped land, became the perfect staging platform for this mission. Through condemnation proceedings, the Navy acquired 158 separate parcels of land totaling 1,680 acres.

One of the main parcels of land was leased and later purchased from the family of Thomas Bard, one of the founding fathers of Port Hueneme. The idea for building the port was the direct result of a coastal exploration by Bard in 1870 during which he rediscovered the Hueneme Submarine Canyon, a trench in the sea bottom a few hundred yards off the coast at the port entrance. In 1871, Bard constructed a 1,500-foot pier to transport goods between the coast and ships offshore and later laid out the town of Port Hueneme.

In 1938, Richard Bard (son of Thomas Bard) and local executives organized the Oxnard Harbor District to develop a deep commercial seaport. On January 24, 1939, operations began to dredge out the harbor. By June 1940, a large deepwater harbor had been dredged out of the low-lying beach and an offshore jetty had been built.

On March 12, 1942, the Navy shipped out scores of trucks with vital materiel and equipment to be used for the construction of the new base. Preliminary work began in April 1942. To speed construction, thousands of Quonset huts were sent from Davisville, Rhode Island, and erected to provide personnel with quarters, warehouses, and shops.

Camp Rousseau, an advance base receiving barracks for transient Seabee personnel, was established on October 23, 1942. The receiving barracks trained over 10,000 personnel in the latest military combat techniques as well as their chosen trade skills. A quarter million men passed through ABD Port Hueneme onto Pacific fronts, and more construction materiel and men were shipped through Port Hueneme than any other port in the United States.

In World War II, the Seabees built the Navy's bases around the world and, with their innumerable construction skills, paved the roads to victory in the Atlantic, Alaskan, and Pacific Theaters. Their accomplishments during the war are legendary, including building over 400 advance bases, 111 major airfields, 441 piers, 2,558 ordnance magazines, hospitals to serve 70,000 patients, and housing for 1.5 million men. Nearly 325,000 men, master artisans, and the most proficient of the nation's skilled workers paved the road to victory for the allies in the Pacific and in Europe. They served on four continents, more than 300 islands, suffered more than 300 combat deaths, and earned more than 2,000 Purple Hearts.

After the war, the workforce cut back to minimal military and civilian personnel and the base almost closed. The Navy, however, decided to keep the $39-million facility operating by incorporating several other activities, tenant commands and organizations. After the war, the Seabees were reduced from 250,000 to 20,000 men and all Seabee activity was concentrated at CBC Port Hueneme, as ranks dwindled to only a few battalions and detachments.

In June 1950, the Seabees were called up to duty again, following the invasion of South Korea by Communist North Korea. The reserve and active-duty force expanded to 14,000 men prior to their landing at Inchon on September 15, 1950. The base shipped construction supplies and equipment and simultaneously trained, equipped, and outfitted new battalions for frontline as well as lateral support in Guam, the Philippines, and Okinawa. A majority of the Navy's construction equipment and supplies used during Korea were directed through Port Hueneme.

The rapid demobilization that followed World War II was not repeated after the Korean Armistice was signed in July 1953. CBC Port Hueneme logistically supported all construction efforts in Southeast Asia and the Pacific during the beginning of the cold war. The most impressive assignment undertaken by Seabees was the construction of Naval Air Station Cubi Point in the Philippines. The Seabees blasted coral to fill in part of Subic Bay, cut a mountain in half for an airfield, moved trees 150 feet tall and 8 feet round, relocated a fishing village, filled in a swampland, and invested 20 million man hours to build this major naval base.

The Seabees had been involved in Civic Action Programs in Southeast Asia since 1954, but efforts heightened in 1963 when the Seabees sent civic action teams and Seabee technical assistance teams into Vietnam and Thailand in an effort to help others help themselves through humanitarian programs. These 13-man teams, referred to as the Navy's Peace Corps, aided rural populations by providing training, technical assistance, and construction work.

The first full Seabee battalion arrived in the Republic of Vietnam in May 1965 to increase military infrastructure construction. The Seabee commitment continued to expand in Southeast Asia until they reached their peak strength of 26,000 men in-country by 1968. They built air bases, port facilities, combat camps, and support facilities in an effort to support the land, air, and sea forces.

When Saddam Hussein and the Army of Iraq invaded Kuwait in August 1990, the Seabees once more answered the call for help. More than 5,000 Seabees built camps, power and waste disposal facilities, and countless miles of roads in the Saudi Arabian desert. In all, the Seabees built 14 galleys capable of feeding 75,000 people, a six-million-square-foot aircraft parking apron, and built 4,750 buildings. They also changed 200 miles of unpaved desert into a four-lane divided road that was the main supply route.

The Seabees have been instrumental in providing support during Operation Enduring Freedom in Afghanistan and Operation Iraqi Freedom by repairing tarmac damage, vehicles, and machinery. They cut steel plating used to enhance tactical vehicle armor, patrolled the streets of Fallujah prior to Iraq's historic democratic elections, built schools and implemented improvements to local village water, electricity, and sanitation facilities.

From the islands of World War II, the beaches of Korea, the jungles of Vietnam, and the deserts of Kuwait, Iraq, and Afghanistan, the Seabees have built entire bases, bulldozed and paved hundreds of thousands of miles of roadway and airstrips, and accomplished a myriad of construction projects.

Navy Seabees deploy around the world to provide construction support for American forces as well as humanitarian assistance and disaster recovery. Using Seabee "Can-Do," they have been directly involved in projects that have assisted millions of people from the Thailand tsunami victims to the refugee camps in Somalia, Bosnia, Vietnam, Kuwait, South America, and Kurdistan. The Seabees at Port Hueneme also support disaster recovery efforts closer to home. During and after the Northridge earthquake, the innumerable local fires, and the crash of Alaska Airlines Flight 261 off the coast of Oxnard in January 2000, the Seabees responded to the call for help. They were also some of the first to respond to Hurricane Katrina to aid their brethren Seabees in Gulfport, Mississippi, and the local victims in New Orleans.

One

THE EARLY YEARS

1875–1942

Congress allocated $22,000 on March 3, 1873, for a lighthouse to mark Point Hueneme. Pictured here in 1937, the Point Hueneme Lighthouse, built in a Craftsman style with a Swiss and Elizabethan influence, consisted of a two-story residence, with a square tower extending an additional story above the dwelling's pitched roof.

Ventura Road (outside Berylwood on the Bard estate) is still dirt and shaded with the hundreds of original eucalyptus trees Thomas Bard had shipped from Australia. This photograph dates to the 1890s.

This aerial image of Point Hueneme shows the farms on the Oxnard Plain before the harbor was constructed in 1939.

The original Bard family home at Berylwood is pictured here prior to 1911, when it was torn down to construct Bard Mansion. The first floor of the Victorian building was built in 1878, and the upper two floors were finished in 1890.

In the 1890s, when this photograph was taken, Thomas Bard was president of Union Oil Company. Large deposits of petroleum were discovered in Ventura County and developed commercially under the Union Oil Company. Bard later served as a U.S. senator from 1900 to 1905. He purchased the land in 1871, developed an interest in horticulture and silviculture, and was personally responsible for the extensive collection of trees on the grounds of Berylwood.

Mollie Bard poses for a c. 1885 photograph in the family home. Mary Beatrice (Mollie) Gerberding married Thomas Bard in San Francisco in 1876. They had eight children—four sons and four daughters—and raised them in Ojai and Port Hueneme.

Berylwood, locally known as Bard Mansion, was completed in 1912 and was the third residence of Senator Bard and his family, pictured here c. 1920. The house remained in use from 1912 until Mollie passed away in 1937. The home remained vacant until the Navy occupied the land in 1942.

The Richard Bard House was built in 1910 as a one-story cottage and enlarged to two stories in 1926. Thomas Bard and his family lived here during the 1911–1912 construction of Bard Mansion. Richard Bard and his new bride, Joan, moved in after their marriage. Now known as Quarters A, the home has housed the commanding officer of CBC Port Hueneme and his family since World War II.

COMPLIMENTS OF
BERYLWOOD STOCK FARM HUENEME CALIF.
THE HOME OF PRINCE AAGGIE OF BERYLWOOD
SEVEN NEAREST DAMS AVERAGE 1120 LBS. BUTTER
THE HIGHEST RECORD BULL IN SERVICE ON THE PACIFIC COAST

The Berylwood Stock Farm and staff are pictured outside the dairy in 1921. The stock farm was home to Prince Aaggie, Bard's prize bull with the highest record of service on the Pacific Coast.

The official ground-breaking ceremony for Port Hueneme Wharf and the Oxnard Harbor District was held on February 4, 1939, with Judge Blackstock as guest speaker and Oxnard attorney Mark Durley serving as master of ceremonies. Richard Bard, destined to be known as the "Father of Port Hueneme," was asked to turn the first shovel.

On January 24, 1939, the San Francisco Bridge Company began operations in the channel to dig out the Hueneme wharf. The idea for building a port was the direct result of a coastal exploration by Thomas Bard in 1867. Bard had learned of a strong offshore underground aquifer flow of fresh water that created a deepwater trough called Hueneme Canyon. The canyon came within 300 feet of the proposed channel, saving millions on dredging costs and keeping the channel free from sediment.

By May 1939, the Standard Dredging Company broke through the sand dunes and deposits of sediment that lay between the harbor and open sea. In 1938, local farmers created the Oxnard Harbor District with a bond issue of $1.75 million, started dredging, and built a transit shed and a wharf. The harbor officially completed construction by July 4, 1940, but the opening ceremonies were not held until July 5 and 6.

This February 1942 aerial photograph of the new port was taken by the Bureau of Aeronautics during their land survey to discover a suitable location for a naval shore establishment on the Pacific Ocean.

This Bureau of Aeronautics land survey image of the Oxnard Canneries on the Hueneme wharf, dated March 20, 1942, shows a northerly view of Lease No. 9, Parcel 23 of land taken under condemnation for World War II.

The main warehouse building at the Hueneme wharf is pictured here on March 21, 1942.

Two

ADVANCE BASE DEPOT
PORT HUENEME
1942–1946

On September 9, 1943, a unit of the 92nd Naval Construction Battalion marches in formation at the Third War Loan Drive parade in Ventura, California. Over 1,300 Seabees from Camp Rousseau took part in the opening of the Ventura County Third War Loan parade and the dedication of a War Bond Victory House. The 129-piece band composed of men from Ship's Company and the 90th, 91st, 92nd, 95th, and 99th Naval Construction Battalions led the Seabees through Ventura.

Pictured here in February 1944, the Seabee Family Tablet remained the logo for Adavance Base Depot (ABD) Port Hueneme during World War II. Carpenter's Mate 1st Class Frank Nagy created the tablet to represent a Seabee in full field equipment with his wife and child at the poignant moment of parting for war.

Construction expansion at Camp Rousseau, pictured here in November 1943, was necessary in order to house five more battalions. The expansion included the construction of 190 barracks, each housing 32 men on the double-tier plan for enlisted personnel as well as 15 officer's quarters, 25 administration and storage buildings, 25 heads, 5 brigs, and 5 sick bays. The growth provided a 50 percent growth in the personnel stationed at the base.

During World War II, equipment and materiel for the Seabees stretched as far as the eye could see. An average of 5,000 railway cars a month and daily truck convoys brought supplies into the base. More navy cargo was shipped from the Port Hueneme harbor to the Pacific and Alaskan areas than from any other port in the United States.

Advance base equipment is set for shipment to the Pacific Theater in this photograph taken August 18, 1943. Seabee construction of new sea and air bases was a vital factor in the success of American naval operations in World War II. Advance base equipment consisted of standard items available in the domestic market and special items developed to speed and ease construction in forward areas.

Advance Base Receiving Barracks Camp Rousseau housing and training area is pictured in October 1945. Ranges, obstacle courses, weapons schools and administration buildings were included in the training area. Quonset huts were utilized to house Marine and Seabee instructors. The administration buildings included executive offices, a galley and dining hall, officers' quarters, a recreation room, and a library.

Advance Base Depot Port Hueneme and Advance Base Receiving Barracks Camp Rousseau face the Pacific Ocean in this photograph taken September 22, 1944. In the lower left is the city of Port Hueneme, the original Bard Estate (before the Navy purchased the last of the land), and undeveloped land on Ventura Road. The beach in the upper right-hand corner, used during World War II for ordnance practice, is now the Channel Islands Harbor.

Pictured June 13, 1944, this is the private entrance to Bard Estate, opening onto Ventura Road.

The Bard barn and milk house, built in 1918, remained a prominent feature of the local countryside until the barn was demolished in 1951. During World War II, the barn was utilized as a dance hall for military personnel.

A World War II Quonset hut is being skinned in this September 27, 1943, photograph. These prefabricated, lightweight buildings were easily shipped, constructed, and modified to create any size and style of building anywhere in the world. Over 100,000 were created as part of the war effort. After the war, they were adapted for administration buildings and housing units, many of which are still in use on the base today.

The main objective of Ship's Service No. 1, also known as Dry Canteen No. 1, was to supply base servicemen with needed provisions at the lowest possible price, as pictured here in 1944. Five large Quonset huts, each 100 feet in length, contained the dry canteen, wet canteen, cobbler shop, barber and tailor shop, and the dry cleaning. Located on Twenty-third Avenue, near Gate 21 (now the Sunkist Gate), the combined ship's services sales volume remained at nearly $10,000 a day from May 1943 until 1946.

Pictured September 10, 1945, the commissary operated similar to a supermarket and offered produce, household goods, and fresh products but did not carry candy, tobacco products, or alcoholic beverages, which regulations did not permit the store to carry.

The interior of the original naval commissary at ABD Port Hueneme is pictured September 10, 1945. Naval military personnel and their dependents enjoy the privilege of purchasing groceries at approximately 30 percent off of regular retail prices. This savings allows military personnel and their families to stretch their service pay.

The battalion post offices and advance base depot post office were located in two elephant Quonset huts and are pictured here on March 2, 1944. The staff of 75 sold approximately $30,000 in stamps each month and sent $100,000 in money orders.

Battalion post offices handled a certain number of battalions, construction battalion maintenance units, and special drafts at each window. The postal clerks handled 75,000 incoming letters, 22,000 parcels and papers, and 45,000 outgoing letters every 24 hours. This photograph was taken March 2, 1944.

The barbershop opened in May 1943, with 24 chairs and "modern" apparatus. Located inside the barbershop, the tailoring service had 18 sewing machines and was equipped to handle all types of cutting and stitching required to properly fit GI clothes.

Galley staff at ABD Port Hueneme, pictured here February 1944, planned and prepared over 210,000 meals each week. Approximately 50 cooks, bakers, butchers, and assistants serve over 12,000 men three times a day and create 5,000 box lunches a day for the night watch and sick bay.

Seabees at Camp Rousseau wait in line at chow time in this 1943 photograph. The raw products required in the galley each week to feed the base personnel included 100,000 pounds of beef and 50,000 pounds of pork, bacon, and ham; 216,000 eggs; 560,000 quarts of milk; 21,000 pounds of butter; 35,000 loaves of bread; 130,000 pounds of potatoes; 84,000 pounds of coffee; 42,000 pounds of sugar; and 10,800 quarts of cream.

The 24th Naval Construction Battalion appears at the commanding officer's Saturday morning inspection in this view looking northwardly on October 31, 1942. Officers held inspections at quarters to check for cleanliness of uniform and person. At the Saturday inspection, the men wore their best uniform.

The 24th Naval Construction Battalion, Company A is pictured at bag inspection on October 14, 1942. Division officers conducted the inspections, teaching the men how to keep their clothes clean, tidy, properly marked, and up to the allowance required by the Navy.

Anti-aircraft warfare training was held on the beaches at Point Mugu, pictured here in 1944. Live firing exercises were conducted on the beaches of Oxnard and Port Hueneme. This live fire and searchlight training over the ocean continued until the end of the war.

In this March 16, 1944, photograph, new Seabees begin military training. After packs and guns are issued, the men began what was known as "commando training" by climbing down the landing net in a ship-to-shore move. After running a few feet in the sand, they "disembark" from a dry-land ship's prow and learn how to take their carbine, pack, and themselves over the side.

The 25th Special Naval Construction Battalion practice bayonet thrusts and parries in this photograph taken April 27, 1944. Members of the 25th Special include, from left to right, Machinist's Mate 3rd Class George Webb, Quartermaster 3rd Class Eldon D. Reece, Shipfitter 3rd Class Thomas Murray, Seaman 2nd Class Paul J. Smailis, Seaman 2nd Class Donald J. Stonish, Chief Storekeeper Stephen J. Mokus, Water Tender 1st Class Charles D. Mason, Seaman 2nd Class John H. Murch, Seaman 2nd Class Eugene F. Satowski, Seaman 2nd Class Hugh J. Osbourne, Seaman 2nd Class William J. Fessler, Coxswain Vernon A. Timms, Storekeeper 3rd Class Harry J. Mott Jr., Boatswain's Mate 2nd Class Pete Hill, Boatswain's Mate 1st Class Elisha E. Hendricks, and Chief Electrician's Mate George W. Mithun.

Marine instructors taught grenade training as part of "commando training" for all Seabee battalions and detachments on board the station, as pictured here on March 16, 1944. Included in the training were classroom lessons on the inner workings as well as hands-on experience throwing grenades. Thirty military tactical schools operated within Camp Rousseau, each staffed with instructors well trained and experienced in the latest military combat techniques.

Seabees in military training enter a gas chamber to test out the gas masks issued as part of their overseas equipment in this photograph taken June 29, 1944.

The men pictured here in February 24, 1944, from the 113th Naval Construction Battalion and Construction Battalion Maintenance Unit 574 were but two of the hundreds of Seabee battalions headed for the Aleutian Islands or South Pacific outfitted with overseas gear by Camp Rousseau's Supply Department. Standing behind the counter are, from left to right, Storekeeper 2nd Class Ted Labeck, Storekeeper 3rd Class L. J. Haygood, Seaman 2nd Class D. M. Berke, and Seaman 2nd Class F. L. Olarsch. Receiving their gear, from left to right, are Machinist's Mate 2nd Class E. S. Goode and Chief Commissary Steward G. J. Macluckey.

Republic Films creates tropical Pacific Island "X" on a strip of beach at Point Mugu for the movie *The Fighting Seabees* in this November 10, 1943, photograph. A Republic camera unit shoots the landing scenes on the imaginary island. Construction Battalion Maintenance Unit 549 spent several weeks on location at Iverson's Ranch in the San Fernando Valley and Camp Rousseau as extras for *The Fighting Seabees*. Other Seabee Camps used during the filming of the picture include Camp Peary in Virginia; Camp Endicott in Davisville, Rhode Island; and the Marine base in Camp Pendleton, California.

The Fighting Seabees premiered at Camp Rousseau on January 19, 1944. Approximately 10,000 Camp Rousseau Seabees stood in line for two hours and packed both theaters at all seven performances of Republic Studio's $1.5 million production that starred John Wayne and Susan Hayward.

Perched atop her symbolic throne (a Seabee bulldozer), Susan Hayward was crowned the first queen of the Seabees by the enlisted men of CBMU 515 in November 1943. She was officially welcomed to Camp Rousseau by Capt. H. P. Needham, CEC, USN, and crowned by Fireman 1st Class Clarence Wofford and Machinist's Mate 2nd Class William Murray.

As pictured here in 1944, the Seabees created their own victory garden to offset food shortages as one of their contributions to the war effort. The U.S. government proposed the gardens as a national food-growing effort, similar to the liberty gardens of World War I. Empty lots, school fields, former flower gardens and backyards were cultivated for victory gardens. Even people who had never held a hoe or spade or worked with fertilizers and seeds were raising fruits and vegetables in tiny garden plots.

The Voices of Victory Recording Studio trailer allowed personnel to send their voices home on a record that was mailed gratis anywhere in the United States. Pictured April 6, 1944, above at the table, composing their messages, are, from left to right, Coxswain L. J. Gadman, Carpenter's Mate 2nd Class Michael J. Carroll, Seaman 2nd Class Herschel Lewis, Machinist's Mate 3rd Class William A. Veslely, Machinist's Mate 3rd Class Elmer Cox, Water Tender 1st Class H. V. Hounshell, Shipfitter 1st Class H. W. Paul, Seaman 2nd Class W. D. Allen, and Hospital Apprentice 1st Class Robert Lockett.

The Portable War Bond Office, pictured here June 29, 1944, facilitated an intensive drive on the base to get military and civilian personnel to purchase bonds. The sale of war bonds provided a way in which patriotic views and the spirit of sacrifice could be expressed. The bonds became a primary way for those on the home front to contribute to the national defense and war effort. While the initial goal of the war bond campaign was to finance the war, the positive impact on the morale of Americans was perhaps its greatest accomplishment.

Ship's company men make a beeline for the war bond trailer on payday, July 5, 1944. Selling the bonds from the trailer is petty officer Bob Sterling. The buyers are, from left to right, Petty Officers Sam Haynes, Roy Hutchins, S. L. Knight, Otto Miranda, C. W. Jett, and Andy Karitas. Also pictured are Petty Officers S. Lupo and A. T. Januzzi.

Military and civilian personnel wait to donate blood as part of the Red Cross blood donor program. The Red Cross maintained an office at ABD Port Hueneme that administered to all the needs of military and civilian personnel. At home, millions of volunteers provided comfort and aid to members of the armed forces and their families, served in hospitals suffering from severe shortages of medical staff, produced emergency supplies for war victims, collected scrap, ran victory gardens, and maintained training programs in home nutrition, first aid, and water safety.

BM2c R. O. Babbitt, a member of the Shore Patrol, regulates departure of the late liberty buses in this photograph taken September 14, 1944. The liberty buses transported personnel between Oxnard City Park, Ventura, and Port Hueneme. The bus service enabled men on the base—whether on business, liberty, or leave—to catch a bus every half hour, thereby eliminating the hitchhiking method of traveling between the base and Ventura or Oxnard. The liberty buses also traveled to Los Angeles and Santa Monica every weekend.

Machinist's Mate 3rd Class W. E. Benney cautions military personnel on liberty against hitchhiking in town limits in this photograph taken September 14, 1944. The major purpose of the Shore Patrol was to assist and protect military personnel ashore during off-duty hours. Each member of the patrol was well acquainted with the area and provided directions and general assistance. They also handled authentic complaints of military personnel and protected the servicemen when on liberty.

Launching Navy Week festivities for Ventura County Ace Marine flyers, Navy Seabees, entertainers, and golf idols thrilled 12,000 service members and civilians at the Montalvo Golf Club on October 24, 1943. Bob Hope headlined a group of entertainers that included Don Ameche, George Murphy, and Henry O'Neill at the "Stars and Stripes" Golf Tournament. Sponsored by the Oxnard USO, the event was the climax of the Navy Day program.

Bob Hope and Norah Moran entertain the troops at Camp Rousseau on October 26, 1943. *The Bob Hope Show*, performed before 2,300 wildly cheering Seabees in Theater A, featured some of the greatest names in entertainment, including Veronica Lake, Tony Romano, Randall Niles, Jerry Colonna, Vera Vague, and Frances Langlord.

Pictured on November 9, 1944, the Camp Rousseau marching and dance band played for all base functions and furnished dance and swing music for local USO canteens, the Hollywood Canteen, and civic organizations.

The U.S. Navy Band performs at the Camp Rousseau Bond Drive Show, Ventura, California, in this June 23, 1944, photograph. The audience purchased $50 or higher war bonds for entrance to the show put on by Camp Rousseau musicians at Ventura Junior College. The drive brought in half a million dollars towards the Camp Rousseau Bond Drive. The stage backdrop of huge war bond replicas and the fighting Seabees was designed and constructed by Carpenter's Mate 3rd Class William Hendricks.

The original Thomas Bard barn loft, pictured in 1943, was converted during World War II into a dancehall for base commands, USO canteens, and local civic organizations.

The LION VIII Cats dance band is pictured here in 1944. The base administrative staff realized that training black recruits in segregated units would destroy morale and initiative. To help prevent this from occurring, programs were designed to make their experience both educational and encouraging. Morale-building programs included baseball, basketball and football teams, bands, dance orchestras, choirs, glee clubs, and quartets.

Pictured in 1944 is an "All-Negro Dance" in Auditorium A. All-base dances were held often, but black military personnel were not allowed to attend. The base arranged for separate recreational facilities and entertainment and invited women from local USO and AWVS groups, busing them in from Los Angeles and Santa Barbara.

S2c J. E. Smith of the 12th Special Naval Construction Battalion and Frances Smith of Los Angeles exhibit their talent at jitterbugging at the All-Base Negro Dance held in Auditorium A on September 8, 1943. The Los Angeles Negro USO and AWVS (American Women's Voluntary Service), the Hollywood Canteen, and the Santa Barbara AWVS-USO invited the young women and provided their transportation to the base.

The quartet from the 80th Naval Construction Battalion performs at a war bond drive in this photograph taken June 29, 1944.

The 31st Naval Construction Battalion embarks for a tour in the Pacific Theater on October 2, 1944. The 31st, 62nd, and 133rd Battalions participated in the battle for the Japanese-held volcanic island of Iwo Jima. Marines and Seabees stormed the beaches of Iwo Jima on February 19, 1945, after days of bombardment by offshore U.S. Navy ships.

Here is a muddy Seabee camp on Attu, Alaska, in December 1943. Approximately 2,000 Japanese soldiers attacked and then occupied Attu Island, Alaska, in June 1942, following their unsuccessful attack on Dutch Harbor. The Seabees arrived in May 1943 and, along with the U.S. Army, pushed the Japanese to the other side of the island. The 23rd Naval Construction Battalion arrived first, followed by the 22nd, 66th, and 68th Naval Construction Battalions, and the 8th Special Naval Construction Battalion.

The 68th Naval Construction Battalion lay Marston matting on the East-West runway at Attu, Alaska, as pictured here March 17, 1944. The Allies and Japanese each devised ways to deal with the rough conditions of forward airfields in the Pacific and tropics, that were constantly plagued by torrential rains, drainage problems, and soft or unstable ground. Though rigid enough to bridge over small surface inequalities of the ground, Marston matting, a pierced steel planking used to establish useable surfaces for airfields, was rigid enough to bridge over small surface inequalities of the ground and was used to best effect on stabilized subgrade. This combination provided an adequate semi-permanent runway that was widely used in every theater of operations.

The 38th Naval Construction Battalion base at Adak, Alaska, is pictured during the winter of 1942. Naval installations at Adak were planned to provide fleet support by constructing an air station, net depot, PT-boat base, Marine barracks, and general base facilities on the weather-beaten, uninhabited island.

The 63rd Naval Construction Battalion operated a photographic laboratory on Guadalcanal. Photographer's mates cover news events, ceremonies, and accident investigations and provide images for release to Navy and civilian publications or for use in Navy historical documents. Their work includes portrait photography, photographic copying, aerial photography for mapmaking and reconnaissance, production of training films, and all types of audiovisual work.

In this 1943 image, the 14th Naval Construction Battalion is at work on one of the many bridges constructed by Seabees on Guadalcanal. The principal objective of the first phase of the battle for Guadalcanal included denying the enemy and possessing the airfield that the Japanese had been constructing since early May 1942. Japanese resistance was fierce and persistent, continuing for six months until all remaining troops evacuated on February 8, 1943.

Personnel from the 36th Naval Construction Battalion repair the Piva bomber strip with Marston matting after a severe shelling on Bougainville, Solomon Islands, on March 8, 1944.

Seabees of the 41st Special Naval Construction Battalion are pictured on Hollandia in 1944. The Seabee special performed a vital function of desperately needed stevedoring. The initial battalions were composed of personnel well qualified in cargo handling and ship loading. Later recruits were sent to a stevedore training school at Camp Peary, Camp Endicott, or Camp Rousseau. Each special consisted of one construction platoon, since each battalion was required to erect its own living accommodations.

Members of the 34th Naval Construction Battalion's Servicemen's Christian League are pictured at Kukum, Guadalcanal, British Solomon Islands, on July 30, 1944. Established in November 1942, the 34th was the first black battalion of the Naval Construction Forces. During their 21-month tour of duty on the Solomons, they built airfields, roads, warehouses, hospitals, and other military installations under almost constant enemy fire.

A member of a Naval Combat Demolition Unit on Island-X gears up in deepwater diving apparatus on December 11, 1943. Naval Combat Demolition Units were among the first to go ashore on D-Day. Along with the Army Corps of Engineers, their crucial task remained to destroy the steel and concrete barriers the Germans had constructed along the beaches of Normandy to forestall any Allied amphibious landings.

A photographer was on hand to document this demonstration of full deepwater diving gear for the Naval Combat Demolition Units on December 11, 1943. Natural and enemy-made obstacles had to be cleared in every major invasion during the war. The personnel received intensive training in blowing channels through sandbars with an explosive hose, working with rubber boats to place explosive charges on underwater obstacles, and using swim-diving apparatus.

Once the maximum feasible use had been made of existing defense facilities in England during World War II, it was necessary to provide facilities of Quonset huts and tents. Suitable alterations were made to existing hotels and large homes to fit them to hospital and dispensary purposes. In 1943, the Seabees attached a Quonset hut annex to a country manor to create St. Michael's Hospital, a temporary facility in Falmouth, England.

A Rhino ferry manned by Seabees from the 111th Naval Construction Battalion provided a secret weapon of enormous value in establishing beachheads on the French coast during the D-Day invasion of June 6, 1944. Rhinos were use to unload equipment from an LST using a special unloading ramp placed opposite the engines. The Rhinos used 186 pontoons to assemble into a huge ferry.

The 111th Naval Construction Battalion land at Omaha Beach on June 6, 1944, before the Mulberry was installed. Ships unload on the beach while barrage balloons hover overhead. The Seabees constructed and operated camps for naval personnel behind the invasion beaches. On D-Day plus six, work began on a beach camp designed to accommodate 6,000 men.

The Seabees of the 111th Naval Construction Battalion give thanks on D-Day plus 12 (June 18, 1944). Navy chaplains have served around the world with Seabee battalions since their inception in 1942, conducting regular services using any available area: a ship's deck, an apple orchard, a hand-cut hole in a Pacific-island jungle, or a makeshift tent for a church. They used a jeep, a packing case, or an ammunition box for an altar, perhaps a helmet for a yarmulke, the top of a mess kit for a paten, and even a canteen cup for a chalice.

The 302nd Naval Construction Battalion Seabee Barge Crew poses with captured Japanese flag and gun aboard a beached three-by-seven-foot pontoon barge on Saipan, Mariana Islands, in June 1944. Pictured, from left to right, are Seaman 2nd Class Charles W. Barrett, Boatswain's Mate 2nd Class Wally Jalloway, Seaman 2nd Class Robert Fisher, Chief Carpenter's Mate (AA) Ragnar Farnum, Machinist's Mate 3rd Class Paul Rhoades, and Shipfitter 3rd Class Carl Jones.

The 302nd Naval Construction Battalion build a pontoon causeway pier at Blue Beach No. 2 on D-Day plus nine in Saipan, Mariana Islands.

The 92nd Naval Construction Battalion base camp on Tinian, Mariana Islands, is pictured here in 1944. The 4th Marine Division accompanied by the 18th and 121st Naval Construction Battalions moved across the strait from Saipan to Tinian on July 24, 1944. The north and west airfields were enlarged to handle B-29 Superfortress landings. The *Enola Gay*, carrying the atomic bomb "Little Boy," left Tinian's North Field for Hiroshima on August 6, 1945.

In this photograph taken April 30, 1945, Seabees from the 62nd Naval Construction Battalion work around the clock to build the B-29 air strip on Iwo Jima. The tower in the foreground and others to the right were built for floodlights for night work. Three Seabee battalions landed with the Marine invasion forces on February 14, 1945. The capture of Iwo Jima, halfway between Saipan and Okinawa, provided a significant site for the development of airfields to support the operations of fighters escorting B-29s in their missions over Japan and an emergency landing field for aircraft returning from raids.

Seabees put their multipurpose pontoons to a new use shortly after the Okinawa operation in the construction of this seaplane repair platform, photographed August 30, 1945. The pontoon platform was used to bring damaged and powerless planes to repair bases. The propulsion barge was built of the same five-by-seven-foot steel boxes used to construct causeways and supply barges. The barge pictured here is transporting an ABM patrol bomber.

In July 1945, the 71st Naval Construction Battalion rebuilt native roads that were unable to handle the heavy traffic of combat operations. Heavy rains and military and construction equipment caused the roads to deteriorate and forced the commanding general of the Fifth Army to restrict all traffic to supply provisions. Work on other construction projects halted, and construction troops concentrated on road maintenance.

Three

HOME OF THE
PACIFIC SEABEES
1946–1962

Passersby on Ventura Road were pleasantly surprised the morning of September 17, 1954, when a new sign appeared designated CBC Port Hueneme as the "Home of the Pacific Seabees, MCBs Two, Three, Five, Nine and Eleven." The sign was designed and painted by Irving "Pinky" Pickney of the base's Public Works Paint Shop.

The new Thomas barracks, pictured under construction on August 4, 1953, represented the first permanent construction at CBC Port Hueneme since the base was established as a permanent naval installation. The barracks provided standard berthing and messing facilities for military personnel attached to the command. The project consisted of 10 three-story barracks, a substance building, a bakery, and a central boiler plant.

Wherry Housing and Thomas Barracks are pictured on April 25, 1954. In the spring of 1949, a tent city was erected as an emergency measure to house Mobile Construction Battalions (MCB) while homeported until permanent barracks could be built. The Thomas Barracks, dedicated to Capt. Robert E. Thomas, CEC, who was killed in an air disaster while en route from Honolulu to the United States in January 1953, were the first permanent barracks built on the base to house MCB personnel.

Pictured on March 18, 1953, Ventura Road was a two-lane paved road near Bard Mansion. Thomas Bard imported thousands of trees from around the world to adorn his family garden and arboretum in Port Hueneme.

The War Assets Corporation, established on February 19, 1944, was charged with auctioning all war assets to the public and supervising disposition of surplus war property. Its building was located at the entrance to Gate 5.

One of the largest auctions of used and unused naval equipment ever offered for sale was held December 7 and 8, 1954. Materiel valued at $8.1 million was auctioned, including automotive, construction, and industrial equipment as well as spare parts. Among the countless items being offered were 650 trucks, 54 cranes, rock crushers, scrapers, air compressors, pole trailer, cranes, graders, bucket loaders, winches, pumps, and generators. This photograph was taken on February 20, 1954.

Pictured May 1950, this pickling unit was used to clean Pacific Theater–area rollback materiel. In May 1949, Advance Base Depot Port Hueneme received all of the Pacific Ocean surplus construction materiel. A cleaning and preservation program was implemented to remove the iron oxide accumulated through lack of proper storage. The pickling unit, created by Supply and Fiscal and the Public Works Department personnel, consisted of eight tanks using degreasers, paint strippers, sulphuric acid, neutralizers, rust preventatives, and water.

The December 1954 Seabee Center auction was the largest auction in history. Some 282 successful auction bidders began removing construction and industrial equipment as well as hundreds of spare parts from the base on Pearl Harbor Day. Sales reached $3.245 million—a 40 percent sales figure. More than 5,000 people attended, with the largest single sale being a $70,000 rock crusher.

Several female office staff members are pictured in November 1960 at CBC Port Hueneme administration building. The first-annual Federal Woman's Award, announced by civil service commissioner Barbara Bates Gunderson, honored six outstanding career women. The award brought deserved recognition to able women in government service and recruited young women with high potential who might otherwise be unaware of the opportunities available to them.

Gate 1, now the Pleasant Valley Gate, is pictured on January 25, 1954, after the installation of the guardhouse and perimeter fence. A security office, built simultaneously outside Gate 1, allowed military and civilian personnel to obtain visitor passes and permits to drive their automobiles on board the base.

Gate 21 along Ventura Road is pictured December 1, 1956, after the new guard building, the EM Club, and the fence line were completed. Gate 21, now the Sunkist Gate, becomes Twenty-third Avenue once inside the gate.

Staff members of the Naval Civil Engineer Laboratory (NCEL) pose outside the administration and laboratory building in the 1950s. The NCEL was unique within the Navy's research and development community. Their principal mission within the Navy was to support the naval shore and harbor facilities, ocean engineering, advance base and amphibious facilities, environmental protection, and energy conservation and power generation through research, development, technology, and engineering.

The original Civil Engineer Corps Officers School (CECOS) building, pictured here on July 25, 1961, was located near Bard Mansion. CECOS relocated from Davisville, Rhode Island, to refurbished World War II barracks at Port Hueneme in September 1947. In 1991, they moved to Moreell Hall, a technologically state-of-the-art building across the courtyard from the barracks. The mission of CECOS is to provide courses of instruction for CEC officers and senior enlisted personnel, through which they may become acquainted with the specialized administrative and technical engineering necessary to equip them for duty in naval billets.

This base all-hands meeting was held at the Needham Theater in 1956. In October 1953, after 11 years, Theater A was renamed the Needham Theater in honor of then commodore H. P. Needham. Commodore Needham spent two tours at Port Hueneme, from December 1942 to August 1944, as officer-in-charge, Advance Base Receiving Barracks, and from 1947 to 1951 as commanding officer, CBC Port Hueneme.

U.S. Naval Schools Construction (NAVSCON) headquarters, the White House, is pictured in 1960, after a facelift. NAVSCON was responsible for the training required to qualify enlisted personnel in the performance of their specialized rates. The school trained qualified men for the trades of builder, construction electrician mate, utilitiesman, draftsman, mechanic, driver, and surveyor.

NAVSCON Company C won "Cock of the Walk" for having the best company in the department and the best appearance at a military review on June 7, 1960. Here Cdr. LaVern Pyles, commanding officer of NAVSCON, awards the company with the bird.

The galley crew shows the amount of food and supplies necessary to feed the Seabees at the center for one day and the men needed to prepare it in the galley. In the foreground of this January 19, 1954, image are commissary officer Lt. Junior Grade C. E. Reed, G. V. Crosby Jr., R. F. Daly, and L. E. Bowers.

Religious services and counsel were provided for military personnel and their dependents by two chaplains from the Protestant and Catholic faiths. Protestant and Catholic services were held regularly, and Jewish services were held every Friday at the Jewish Community Center in Ventura. The CBC Port Hueneme chapel is pictured here on August 23, 1954.

This 1956 view of CBC Port Hueneme Hospital looks toward Gate 21 (Sunkist Gate). The base hospital was constructed in 1943 with 175 beds, 5 general wards, and 2 isolation wards. Built on the site of the current NEX, the hospital contained all state-of-the-art medical and dental equipment with an operating suite for general surgery, a prosthetic dental laboratory, an eight-chair dental office, treatment rooms, X-ray department, laboratory, and pharmacy.

Construction of the new service station on the corner of Easy Street and Twenty-third Avenue, operated by Ship's Service Department, was completed in October 1953. The station offered gasoline (both regular and Ethyl) and motor oil from Monday to Friday, 0830 to 1730. Construction of the station was begun by a civilian contractor and completed by NAVSCON students as part of their training.

The original Seabee Museum opened in 1947 as a monument to the fighting spirit of the Seabees who fought in World War II. Captain Fink, base commanding officer, and Commander Knightly, base executive officer, conceived the idea and began a quest for materials in the fall of 1945. The original building at Pacific Road and Tenth Avenue displayed relief maps, enemy equipment that had been captured, replicas of outstanding Seabee events, and battalion mementos. This photograph was taken on August 15, 1950.

The most noteworthy display in the Seabee Museum was a mural of Seabees at work depicting their famous "Can-Do" spirit, pictured here August 15, 1950. The museum contained exhibits of model equipment used and islands visited by the Seabees, battalion plaques and cruise books, and photographs from various battalion combat sites.

Lt(jg) D. A. Morton poses on the Japanese Zero outside the Seabee Museum on September 23, 1956. The Zero was added to the artifacts at the Seabee Museum in September 1947. It arrived in Oakland on the USS *Arnev* (AKS-56) from Guam and was sent to the base aboard the USS *Andromeda* (AKA-15). Spare parts were also sent along to repair the aircraft.

Seabees on the base celebrated their 18th birthday on March 5, 1960, by inviting the public to an afternoon open house in the style of a three-ring circus, topped off by a grand ball in the evening. Approximately 2,300 visitors witnessed the Seabees in action as they constructed a camp complete with roads and airfield, took up arms to ward off simulated enemy attack, and rescued an injured man atop a light pole. The exhibition trailer displayed all types of Antarctic, advance base, and construction materiel used by Seabees all over the world.

Personnel and military family members at CBC Port Hueneme were quite astonished to wake up and see the station blanketed with snow on January 11, 1949. This Quonset hut in the Homoja area shows the rare snow near the beaches of sunny Southern California. The Homoja housing area consisted of 20-by-48-foot Quonset huts that had been converted into two-family apartments.

The CBC Port Hueneme motor pool was photographed on January 11, 1949, after snow fell and blanketed the streets, houses, and construction equipment.

The Seabees celebrate their fifth anniversary with a gala dance at Port Hueneme in this March 1947 photograph. The "Missouri Waltz" was being played as the celebrants danced in rhythm with the U.S. Navy Training and Distribution Center Band at Theater B. Enlisted men and officers shared equally in the celebration. The original Seabee anniversary was December 28 but the celebration was changed to March 5 in 1947.

Beauty queen Judy Garrett of Ojai is surrounded by her court following the Miss Ventura County beauty contest. Her coronation was held at CBC Port Hueneme's Warfield Gymnasium on June 30, 1956. Pictured, from left to right, are Naomi Perate, Sandra Knocke, Judy Garrett, LaVonne Riley, and Jo Ann Silva.

The third-annual Point Mugu–Port Hueneme Navy Relief Rodeo and Carnival, held June 5–7, 1959, raised money for the Navy Relief Fund. The Navy Relief Society (now the Navy-Marine Corps Relief Society) provides, in partnership with the Navy and Marine Corps, financial, educational, and other assistance to members of the Naval Services of the United States, eligible family members, and survivors in need.

In 1960, the Navy Relief Fund Raising Campaign launched the first Space Fair, the predecessor to the Point Mugu Air Show. The Space Fair featured the Blue Angels and exhibits by many of the county's leading missile and aircraft manufacturers, displays in Navy missiles and jet aircraft, and a carnival midway.

Seabees march on parade along with the U.S. Naval Schools Construction (NAVSCON) float during the early 1950s.

At the Seabee Parade, held on March 11, 1961, Mobile Construction Battalion 10's float dedicated to the Mercury Space Program was entitled "Conceived in War, Dedicated to Peace." A parade through downtown Oxnard and the Seabee Ball at the Warfield Gym climaxed a weeklong celebration of the 19th birthday of the Seabees and the 94th anniversary of the Civil Engineer Corps.

Students attend a refrigeration class during the Utilitiesmen A (Basic UT school) program, as pictured here on March 19, 1951. Utilitiesmen are involved with plumbing, heating, water and distribution systems, sewage collection, and disposal facilities at Navy Shore establishments around the world.

In this 1951 photograph, a chief electrician demonstrates an electrical installation to construction electrician mates of Class A (Basic CE School) by stringing power distribution lines on poles approximately eight feet high. In 1948, the original Seabee training command, Training and Distribution Center (TADCEN), changed to the U.S. Naval School Construction (NAVSCON). The new school realigned their technical training program with eight 12-week-course programs.

In this March 19, 1951, photograph, students from the Builders School assemble a bridge as part of A School training. Immediately following World War II, the Navy Department restructured Seabee ratings to general service ratings. In June 1947, in order to prepare for the new rating structure, the Bureau of Personnel directed U.S. Navy Schools, Construction, Port Hueneme to condense the 19 schools in operation into 8 to cover the new Seabee ratings of Surveyor, Driver, Mechanic, Utilitiesmen, Builder, Steelworker, Construction Electrician Mate, and Draftsman.

Members of Naval Mobile Construction Battalion 9 are pictured on December 1, 1953, during carbine class. The military training program at Construction Battalion Base Unit (CBBU) included undergoing a course designed to familiarize the students with the use of the M-1 carbine rifle.

Eighty-four men from the Oxnard Air Base take the 45-caliber pistol familiarization course at the Construction Battalion Center Military Training Area Range in this photograph taken June 4, 1953. All personnel (both officers and enlisted men) qualified by attaining a minimum score of 184 out of a possible 400.

In 1958, Naval Mobile Construction Battalion 9 made a four-mile march with packs to a bivouac area in the Los Padres National Park.

The Construction Battalion Base Unit (CBBU) Obstacle Course, pictured on September 17, 1960, was organized on the base in May 1952 under the command of the commander in chief of the Pacific Fleet. The unit was responsible for the administration and training of Seabee personnel awaiting assignment to permanent stations. Two instructors, Equipment Operator 2nd Class J. H. Adams (above) and Equipment Operator 3rd Class D. L. Hotz show off the newly installed obstacle course built in accordance with the army field manual.

Naval Mobile Construction Battalion 11 personnel run the obstacle course as part of their military training on September 22, 1960. The unit's new obstacle course was designed to give Seabees a thorough workout. The course had 20 different obstacles and provided supplementary training. It included such obstacles as the island hopper, run and jump, confidence jump, skyscraper, monkey climb, reverse climb, and "dirty name."

Pictured in March 1961, a U.S Naval Schools Construction (NAVSCON) instructor trains Seabees in M1 Rifle stripping. The instructors provided the necessary guidance for conducting basic and automatic rifle marksmanship, insuring proper weaponry maintenance, and applying correct techniques of rifle marksmanship when engaged in combat.

A U.S. Navy truck assists at the Malibu fire of December 31, 1956. The Malibu fire was brought under control after five days of around-the-clock firefighting by civilian firefighters and 600 volunteer Seabees from MCB's 3, 9, and 11. The blaze covered 40,000 acres from Point Dume (near Zuma Beach) to Lake Sherwood and destroyed 75 homes.

Seabees fight a local fire in the Señor Canyon district, where the fire spread from Ojai. Seconds after this picture was taken, a shift in the wind caused the trees and shrubs in the foreground to be completely engulfed in flames. The men escaped with a few singed eyebrows and hair.

Mobile Construction Battalion 9 departs for Alaska in March 1958. The battalion's approximately 450 officers and men left aboard the USNS *Funston*. The main body of the battalion was located at Kodiak, Alaska, and Detachment Alfa was stationed at Adak in the Aleutian Islands.

Naval Mobile Construction Battalion 5 unloads equipment used during Operation Greenlight in this July 25, 1961, photograph. Operation Greenlight was created with the objective of determining the capability of naval construction forces to support a Marine Aircraft Group in the field under advance base conditions and to test the suitability of functional components.

Amphibious Construction Battalion One (ACB-1) land with the United Nations forces at Inchon, South Korea, in September 1950. The tide at Inchon Harbor rises and falls rapidly, making it dangerous to handle the pontoon causeways. Battling 30-foot tides and a swift current while under continuous enemy fire, they positioned pontoon causeways within hours of the first beach assault. The Seabees also built airstrips, cleared mined tunnels, repairing ships, and stole three locomotives to transport heavy equipment to advancing forces.

In June 1951, Seabees of CBMU 1804 created an emergency airfield on the Island of Yo in Wonsan Harbor off the east coast of North Korea under the code name Operation Crippled Chick. The airfield was critical for aircraft hit by enemy fire that had little choice but to ditch at sea or attempt to land in enemy territory. Working under constant enemy artillery bombardment, the Seabees completed the 2,400-foot airstrip in 16 days.

Mobile Construction Battalion 2 on the carrier pier get a pile driver in position at Cubi Point, Philippines, on February 18, 1953. MCB 2 arrived in Cubi Point in June 1952 to join MCBs 3 and 5 with construction of the naval air station at Subic Bay. The land was given to the United States by treaty with the Republic of the Philippines.

A grading operation on the first 200 feet of road from Cubi Point to Camayon Point is underway in this photograph from November 25, 1953. A clearing 100 feet wide was bulldozed and blasted through dense jungle timber prior to grading. A 24-foot-wide paved road connected the ammunition facilities at Camayon Point with Naval Air Station Cubi Point.

Beginning in 1955, Seabees began deploying to Antarctica as part of Operation Deep Freeze to build and expand scientific bases. Seabees built airstrips, the continent's first nuclear power plant, snow-compacted roads, underground storage, laboratories, living quarters, and several permanent scientific bases. The Seabees were part of Antarctic construction projects until 1993.

Here a group of Seabees compact snow for roads to connect the numerous scientific bases throughout the frozen continent.

Mobile Construction Battalions 2, 3, and 5 return from Cubi Point to find their families waiting on the dock for their return on July 10, 1953. MCBs 3 and 5 left Port Hueneme for the Philippines in October and November 1951. MCB 2 included the short-timers who had left in June 1952, 13 months earlier, to join their Seabee comrades in Cubi Point. The Seabees were sent to the Philippines to create a new naval installation at Cubi Point.

Four

HELPING OTHERS
HELP THEMSELVES
1962–1972

Personnel from U.S. Naval Schools Construction, the 31st Naval Construction Regiment, and Mobile Construction Battalions 3, 10 and 11 participate in the Open House parade to celebrate the 100th anniversary of the Civil Engineer Corps and the 25th birthday of the Seabees on March 4, 1967.

This March 4, 1967, photograph shows the presentation of battalion colors at a pass in review to celebrate the anniversary of the Seabees. The battalion honor guard stands at attention during the regimental parade and pass in review held at CBC Port Hueneme. Over 2,000 troops participated in the anniversary pass in review.

Navy Civil Engineer Laboratory (NCEL) divers display a new sign for the NCEL Seabee Diver Locker in March 1968. The NCEL Seabee Divers were the precursors to the Underwater Construction Teams (UCT) that now provide construction, inspection, repair, and maintenance of ocean facilities in support of Navy and Marine Corps operations. Pictured, from left to right, are Capt. L. N. Saunders Jr., NCEL commanding officer and director; Capt. A. F. Dill, executive officer; Irving Pinckney, CBC Public Works, creator of the sign; Lt. Junior Grade B. W. Savant, assistant to the executive officer; and Charles Waddell, Seabee diver.

Fire destroyed the Needham Theater in the early morning hours of July 10, 1967, causing approximately $100,000 in damage. Two engines from the CBC fire department and additional engines from Port Hueneme, Oxnard, and El Rio worked to save the 25-year-old wooden structure. The building was demolished and a new theater constructed on the same site. Movies screened at the Warfield Gymnasium until the new Needham Theater finished construction.

The Disaster Recovery Training Department trained construction battalions in recovery from the effects of natural disasters and from nuclear, biological, and chemical warfare actions. The program included individual, classroom, and team training and a full scale, integrated recovery of an advance base from the effects of a realistically simulated nuclear, biological, and chemical attack. This photograph was taken in 1966.

The 31st Naval Construction Regiment personnel created a model Vietnamese village as part of the open house to celebrate the 100th anniversary of the Civil Engineer Corps and 25th birthday of the Seabees on March 4, 1967. The program featured an attack on the village by the Viet Cong, a concert by the Twenty-nine Palms Marine Corps Band, and a drill team exhibition by men of Mobile Construction Battalion 11.

Attractions in the 31st Naval Construction Regiment's exhibition tent at the 1967 open house included static displays by the regiment and by U.S. Naval Schools Construction, Naval Civil Engineer Laboratory, and DRTD. The open house drew more than 20,000 visitors.

The Seabee teams usually consisted of a junior Civil Engineer Corps officer, 11 construction men, and a hospital corpsman. The teams constructed roads, schools, orphanages, public utilities, built dams, and taught construction skills in rural areas. They were assigned to the Department of State, U.S. Special Services and in support of foreign countries. The teams proved exceptionally efficient in rural development programs and earned reverence as the "Navy's Peace Corps."

JOIN THE SEABEES — SERVE THE WORLD

!ENLIST NOW!

SEABEES
"CAN DO!"

!TOP RATES FOR TOP MEN!

PETTY OFFICER

CONSTRUCTION BATTALIONS

UNITED STATES NAVY

This 1966 enlistment poster was used during the 25th birthday of the Seabees. Beginning in 1963, the United States military buildup in South Vietnam slated the Seabees to play an important and historic role in the growing conflict. By the autumn of 1968, the Naval Construction Force had grown to 26,000 men serving in 21 full-strength battalions, 2 Construction Battalion Maintenance Units, and 2 Amphibious Construction Battalions.

This is the 31st Naval Construction Regiment Seabee Team Training sign outside their administration building. The 31st NCR was assigned the specialized military training of the Seabee teams. The teams were designed specifically for deployment under State Department sponsorship to underdeveloped and emerging nations to help the rural denizens in the fundamentals of modern construction.

Naval Mobile Construction Battalion 11 is pictured in 1966 during grenade training at Camp Pendleton. The two primary objectives of the hand-grenade training program were to develop proficiency in grenade throwing and overcome any fear of handling explosives. Seabees were taught defensive techniques and tactics as well as familiarization with various types of grenades, land mines, flares, and booby traps used by the Viet Cong.

Naval Mobile Construction Battalion 3 personnel began a strenuous program of military training at Camp Pendleton. To ensure maximum military readiness, the intensive program was provided during the final two weeks of June 1968, prior to their departure to Da Nang, Vietnam.

In this November 1966 photograph, an instructor teaches a Seabee from Mobile Construction Battalion 8 how to use the 3.5 rocket launcher during FEX at Camp Pendleton. The 3.5 rocket launcher was used to ignite the rocket propellant and give it initial direction in flight. It provided increased firepower against targets ranging from personnel to heavy tanks.

Instruction on the many different types of mines and disarming procedures was given to the Seabees of Mobile Construction Battalion Eight during their weeklong training at Camp Pendleton, November 1966. The concealed explosive charge was placed in an area where it could be detonated by contact with personnel or vehicles. The mine was buried with the fuse pressure plate just above the ground surface and detonated when pressed down.

A Viet Cong rocket launcher booby trap was used in training at Camp Pendleton, as pictured here November 1966. Improvised booby traps were often attached to an object that could be used or that had a souvenir appeal. The Viet Cong manufactured devices that exploded when touched, picked up, or operated.

Pictured in November 1966, Seabees undergo tactical march training as part of Field Exercises (FEX) at Camp Pendleton. Tactical troop movements are made under combat conditions when troops and equipment are organized, loaded, and transported according to their tactical mission. Tactical march training taught Seabees the threat of enemy action that might include ambushes, land mines, enemy air attacks, and full-fledged attacks by enemy forces as well as security against the action.

Marine instructors train the Seabees of Mobile Construction Battalion 8 in the correct use of the M-14 rifle in November 1966. While construction remained the main goal of the Naval Construction Forces, another mission continued to be the defense of these construction projects when needed.

With USAID and other nongovernmental agencies such as CARE providing the goods, the Seabee Teams were able to distribute hundreds of pounds of foodstuffs and clothing to local Vietnamese villagers, including orphanages and hospitals. In 1966, USAID decided that the Seabee Team program in-country was such a success that they would sponsor more teams to aid in reconstruction, rural development, and distribution of food and clothing products.

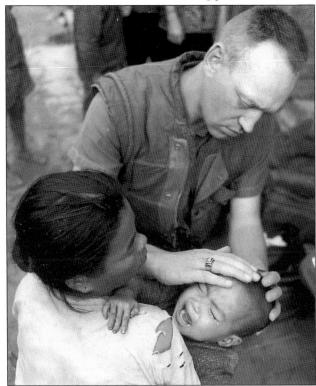

A medical officer from Naval Mobile Construction Battalion 128's Civic Action and Medical Team triages a Vietnamese child in this 1968 photograph. Over 7,000 Vietnamese civilians received regular medical treatment by NMCB-128's hospital corpsman and doctors during their deployment to Vietnam. The Civic Action Companies held sick calls four to six times a week in the Son Thuy I Hamlet, Hoa-Long Village, Sacred Heart Orphanage, and Son Thuy III Hamlet.

During deployment to Vietnam, Naval Mobile Construction Battalion 3's Civic Action Program centered on the Sacred Heart Orphanage in addition to the medical programs held in local villages. The MCB 3's Wives' Club sponsored Operation Diaper and flooded the Sacred Heart Orphanage with clothes and other necessities essential to continued successful operations. This photograph was taken in 1968.

Frequent heavy rains and heavy construction equipment helped to turn the Phu Loc camp in Vietnam into sticky mire, as seen here in September 13, 1967. Mobile Construction Battalion 74 Tango Detail endured mud more than two feet deep during part of their tour in Vietnam.

Personnel of Mobile Construction Battalion 53 lift the assembly jig used in construction of the Liberty Bridge from the water of the Thu-Bon River, 15 miles south of Da Nang.

In 1968, the CBC Port Hueneme Brig sponsored the Seabee Toylift to collect and repair toys for distribution to Vietnamese children at Christmastime. Men confined to the brig reconditioned toys to extend a friendly hand across the sea to children of Southeast Asia and as a means of therapy to rehabilitate prisoners. Deployed Seabee battalions celebrated Christmas in Vietnam with orphans and village children ravaged by war.

A flying crane helicopter creates a large dust storm as it lifts the first of seven Seabee-built observation towers from Mobile Construction Battalion 1's camp near Da Nang.

Seabees assigned to Naval Mobile Construction Battalion 121 lay LAV matting at Camp Campbell in Vietnam. Aluminum plank matting was used to expedite airfield construction by eliminating the time required for placing Portland cement or asphalt pavement. Mats could be laid on the ground or a subsurface of soil-cement. In the hot weather of Southeast Asia, matting teams were required to limit their productivity to a few hours a day because the heat reflections from the aluminum matting sent air temperatures soaring to 125 degrees, thereby making it impossible to touch the panels.

Naval Construction Battalion 10 is pictured in September 1970 erecting the Quang Tri-Highway Bridge.

Personnel of Mobile Construction Battalion 11, Lang Vei Detail, erect a tent at the new campsite in this August 15, 1967, photograph. The tent was used as a workshop for the construction forms during torrential rains that frequent this section of Vietnam.

Assigned to construction and improvement of training facilities of a Civilian Irregular Defense Group (CIDG) camp at Dong Xoai, 55 miles north of Saigon, 9 members of Seabee Team 1104 and 11 other U.S. Army Special Forces personnel were trapped in one of the bloodiest and hardest fought battles of the Vietnamese war. Construction Mechanic 2nd Class Marvin G. Shields and Steel Worker 2nd Class William C. Hooper were killed in action; Lt. Junior Grade Frank A. Peterlin, Chief Equipment Operator Johnny McCully, Builder 1st Class James Brakken, Construction Mechanic 1st Class James Wilson, Utilitiesman 2nd Class Lawrence Eyman, Builder 2nd Class Douglas Mattick, and Hospital Corpsman 2nd Class James Keenan were wounded during the Viet Cong assault.

Pictured here is a Special Forces camp at Dong Xoai after the Viet Cong attack in June 1965. The Special Forces compound supported three Civilian Irregular Defense Groups (CIDG) companies: a Regional Forces company, a small Vietnamese Special Forces detachment, and an armored-car platoon. The Viet Cong attacked close to midnight on June 10, 1965, letting loose a 200-round barrage of 60mm mortars followed by a wave of hundreds of Viet Cong attacking the walls of the compound.

This is a June 1965, rear view of the District Headquarters Building at Dong Xoai. The surviving American troops made their way to the District Headquarters Building but were quickly surrounded by an almost overwhelming Viet Cong. The enemy used flamethrowers, machine guns, recoilless rifles, and small arms against the fortifications. First Lt. Charles Williams and Carpenter's Mate 3rd Class Marvin Shields moved outside the headquarters defenses and successfully destroyed a Viet Cong .30-caliber machine gun position.

The entrance to Camp Shields, Chu Lai, Vietnam, was photographed on September 10, 1967. Camp Shields was dedicated to Marvin Shields, who was fatally wounded during the 14-hour battle with the Viet Cong at Dong Xoai. He was posthumously awarded the Medal of Honor for giving his life to protect his fellow men during the heroic defense of Dong Xoai.

Pictured in 1964, Seabee Technical Assistance Team (STAT) 1103 constructs a road in Nam Pat out of an ox trail. STAT team operations in Thailand began in 1963 at the request of the Royal Thai government. The Seabees were employed in support of Thailand's Accelerated Rural Development program to teach Thai nationals the trades of construction, equipment operations, maintenance, road surveying, and heavy bridge and dam construction.

The Seabees of Mobile Construction Battalion 3 constructed the Nakhon Phanom Airfield, pictured here on December 31, 1963, in the frontier of the Mekong River, Thailand, as a strategic air striking point deep in the interior of Southeast Asia. The battalion convoyed bulldozers, scrapers, and a myriad of construction materiel 500 miles from the port of Bangkok by rail and dirt road to the Laotian border. The Seabees cleared 235 acres of rain forest and laid a 6,000-foot steel-planked runway to create an airbase in the jungle.

Naval Mobile Construction Battalion 5 returns home from Vietnam in 1966. NMCB-5 deployed to Vietnam six times, earning them the dubious honor of the most deployments to Vietnam by a Seabee battalion.

President Nixon greets personnel from Naval Mobile Construction Battalion 3 upon their return from Vietnam in March 1969.

Family members welcome home Mobile Construction Battalion 10 from Chu Lai on December 12, 1965.

Five

HUMANITARIAN ASSISTANCE AND GLOBAL CONFLICT
1973–2000

Naval Mobile Construction Battalion 40 deployed to Bosnia-Herzegovina as part of Task Force 519, IFOR (pictured here in 1996), in support of Operation Joint Endeavor. Throughout the harsh freezing winter and amid the seas of mud, the Seabees of NMCB-40 tactically retrograded U.S. Army base camps, transported materiel, dismantled camps, supported Camp Colt's LSA, removed snow and ice, and completed numerous other missions that spread the personnel throughout the sector.

The City of Port Hueneme erected this sign on Hueneme Road near Ann Avenue to welcome visitors, military personnel, and dependents.

The Needham Theater was named after Capt. Henry P. Needham, the officer in charge of Advance Base Depot Port Hueneme from 1942 to 1944 and commanding officer of CBC Port Hueneme from 1947 to 1952. The original Needham Theater burnt in a fire in 1967 and was rebuilt in 1969.

During the weeklong Advance Base Functional Component and Disaster Recovery Training, Naval Mobile Construction Battalion 5 was tasked with building a 180-foot bridge capable of holding 45 tons of men and equipment, pictured here in February 1986. The 46-man crew labored 10 hours to construct the Bailey Bridge, which was used to move men and equipment across obstacles such as rivers and ravines.

Pictured February 18, 1980, evacuated flood victims from Point Mugu receive cots, sleeping bags, and staples after military personnel and their dependents were taken to CBC Port Hueneme for temporary shelter. Naval Air Station (NAS) Point Mugu flooded after a levee wall and a dike broke on two separate stretches of Calleguas Creek.

Mobile Construction Battalion 5 uses heavy equipment to deepen the channel adjoining Calleguas Creek in this photograph from February 20, 1980. The levee from Calleguas Creek unleashed its devastating stockhold of water that flooded over 80 percent of the NAS Point Mugu. The Seabees used 15,000 sandbags to build a mile-long wall along the Pacific Coast Highway to hold back any increase in the floodwaters.

In response to a request from FEMA to alleviate traffic congestion resulting from the Northridge Earthquake on January 17, 1994, the Seabees of Naval Construction Battalion 40 constructed a train platform and parking area at the Vincent Train Station, located south of Palmdale, California. Construction Battalion Center personnel delivered 32 water tanks to the Simi Valley City Hall and 25 water tanks to Los Angeles County for distribution and assisted FEMA field offices with administrative support.

Naval Mobile Construction Battalion 40 in Ojai, California, assists with widening and deepening the flood control channel of San Antonio and Thatcher Creeks and building various safety berms on the creeks in this November 1986 photograph. The work permitted the creeks to channel more mountain runoff from expected rain to the rivers and ocean.

Railroad flats are loaded with tanks that were used in exercise Team Spirit 1987. The United States and Republic of Korea annually engaged in Team Spirit, a joint operation to improve defense readiness. Equipment transported via convoy, rail cars, and commercial trucks was loaded at Port Hueneme onto the M/V Cape Horne. The base shipped approximately 119,000 square feet of equipment to the Republic of Korea for this exercise.

Seabees from Naval Mobile Construction Battalion 40 load up their equipment for their deployment to Mogadishu, Somalia, in December 1992. Over 50 Seabees from NMCB-40 mobilized to provide construction support to the forces of Operation Restore Hope. They built and repaired schools and orphanages and provided security for nongovernmental organizations so the organizations could distribute food and relief support at nine humanitarian sites.

Pictured in December 1992, members of Naval Mobile Construction Battalion 40's Air Detail drive through the streets of Mogadishu shortly after their arrival in Somalia. The Seabees established base camps at each of the humanitarian relief sites and provided construction support to U.S. and Coalition Forces during Operation Restore Hope. They also built and repaired schools and orphanages for local Somalian children.

Camp Stalder, pictured here in January 1993, was located near the Mogadishu airfield. The Seabee living quarters were on the southwest corner of the airfield and about 200 feet from the edge of the runway, making it noisy and dusty.

Seabees deployed with NMCB-40's Air Detail, sent to Mogadishu, Somalia, as part of Operation Restore Hope, drill and install a well for the local Somalian community in this January 1993 photograph. The Seabees set up lights for the Mogadishu airfield, cleared the streets of old cars, sand, and debris, leveled terrain for fuel bladders, and built numerous water and sewage facilities for the Marine, Army, Air Force, and Allied Forces serving in Somalia.

Pictured in January 1993, a Seabee from NMCB-40's Air Detail aids a local orphanage by installing a floor. Every Seabee put in 12-hour days, with some working nights to complete the various commitments to Allied Forces and the local community.

Naval Mobile Construction Battalion 40's Air Detail is pictured at Camp Stalder, Somalia, in February 1993. Camp Stalder was named in memory of Utilitiesman 2nd Class Jack L. Stalder, who passed away during NMCB-40's prior homeport. Despite the adverse living conditions and extended working hours, the Air Detail upheld the Seabee "Can-Do!" spirit and got the job done in record time.

Naval Mobile Construction Battalion 40 received official tasking to redeploy from Rota, Spain, to Bosnia on September 13, 1996, in support of Operation Joint Endeavor. The battalion deployed 335 Seabees to disassemble and retrograde 14 base camps, complete 19 sustainment projects, maintain and repair the main supply route, and complete construction engineering tasks.

In August 1990, Seabees began deploying to support the First and Second Marine Expeditionary Forces by providing construction, amphibious landing, and underwater construction support. Upon their arrival, the Seabees built critically needed facilities at four Marine Air Combat airfields, a base camp for the Marine Air Wing, Army and Air Force support in Bahrain, and munitions and aviation storage facilities.

During Operations Desert Shield and Desert Storm, the Seabees constructed the largest wartime multi-battalion military complex since Vietnam, nicknamed "Wally World." Pictured here in December 1990, the complex was comprised of six camp modules capable of housing 2,500 men each, with all the berthing, office space, showers, toilet facilities, galley, roads, and parking areas necessary to accommodate the personnel.

Six

THE SEABEES AT PORT HUENEME TODAY
2000–2005

Builder 1st Class Ben Campbell, assigned to Naval Mobile Construction Battalion Three inspects the roadway for improvised explosive devices during exercise Bearing Duel at Fort Hunter-Liggett, California. Campbell served as part of a Seabee Engineering Reconnaissance Team (SERT), which conducts an engineering and security reconnaissance of specified supply routes. The exercise evaluates the Seabees skills for potential future deployments and includes force protection, rapid runway repair, bridge building, and water purification.

Naval Mobile Construction Battalion 40 loads equipment onto an Air Mobility Command (AMC) C-5 "Galaxy" cargo plane at Point Mugu on December 13, 2002. NMCB-40 deployed to Guam in support of disaster relief efforts after Super Typhoon Pongsona passed over the island on December 8, 2002.

Seabees from Naval Mobile Construction Battalion 40 march by during a pass in review at the July 17, 2004, opening ceremony for Seabee Days, an annual event that includes a military parade, carnival, entertainment, and military displays.

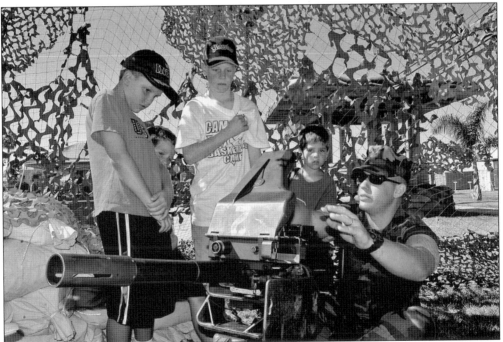

On July 17, 2004, Utilitiesman 2nd Class Nicholas Morgan shows some children how to operate a MK-19, 40mm grenade launcher at one of the Seabee Days displays.

An Underwater Construction Team 2 diver plays tic-tac-toe with the children from inside a water tank at a Seabee Days display on July 17, 2004.

On August 8, 2004, Naval Mobile Construction Battalion 23 members clean 9mm pistols following a qualifications course at Port Hueneme. The class included one day of instruction and one day on the range firing the weapon. NMCB-23 mobilized in support of Operation Iraqi Freedom (OIF) and was training for deployment.

Pictured July 29, 2004, range coaches from the 31st Seabee Readiness Group give a demonstration on the M-136 AT-4 rocket launcher to Naval Mobile Construction Battalion 40 Seabees at Camp Roberts. NMCB-40 Seabees completed training for an upcoming deployment while providing logistic support for deployed Seabee battalions.

In this photograph taken July 9, 2004, Constructionman 1st Class James Richardson steps into the water, one-step behind fellow diver Constructionman 1st Class Steven Hentze. Richardson, Hentze, and other members of Underwater Construction Team 2 were in Diego Garcia to perform maintenance on pier facilities. Underwater Construction Teams provide a capability for construction, inspection, repair, and maintenance of ocean facilities in support of Navy and Marine Corps operations.

Equipment Operator 2nd Class Chris Amescua receives assistance from his dive supervisor in sealing his helmet before conducting training dive operations at Camp Patriot, Kuwait, on March 11, 2003. Petty Officer Amescua is attached to Underwater Construction Team 2, which was forward deployed in support of Operation Enduring Freedom.

Builder Constructionman Daniel Trammel operates a defensive firing position assigned to Naval Mobile Construction Battalion 40 during a "Bailey" bridge project on August 26, 2004. NMCB-40 participated in Exercise Bearing Duel 2004, designed to provide the unit with an opportunity to train in realistic field environments while preparing for contingency operations and future deployments.

Members of Underwater Construction Team 2 (UCT-2) prepare for an explosives training exercise at a remote area in Kuwait on April 16, 2004. UCT-2 was attached to Task Force Mike under Naval Mobile Construction Battalion 74. NMCB- 74, 4, 133, 21, SU-2, and UCT-2 conducted operations in the Central Command Area of Operation in support of Operation Iraqi Freedom.

Ensign Simon R. Mueri, attached to Mobile Construction Battalion 40, relays information to headquarters after completing a Seabee Engineering Reconnaissance Team (SERT) mission during an annual field exercise at Camp Hunter-Liggett on April 12, 2003. Mueri is assigned to Naval Mobile Construction Battalion 40.

Constructionman 2nd Class Joshua Powers takes up a defensive position on a Seabee Engineering Reconnaissance Team (SERT) mission during Exercise Tandem Thrust 2001 at Shoalwater Bay, Australia, on May 1, 2001. The event was a combined United States, Australian, and Canadian military training exercise.

Construction Mechanic 3rd Class Aaron Wolken mans a turret-mounted M-240B machine gun atop a High Mobility Multipurpose Wheeled Vehicle, providing security while Seabees assigned to Naval Mobile Construction Battalion 4 clear debris from the streets of Fallujah, Iraq, on November 17, 2004. NMCB-4 was deployed in support of Operation Al Fajr (New Dawn), an offensive operation designed to eradicate enemy forces within the city of Fallujah in support of continuing security and stabilization operations in the Al Anbar province of Iraq.

Illumination rounds silhouette Navy Reserve Seabees assigned to Naval Mobile Construction Battalion 22 as they adjust the site on the bore of an M-224 60mm mortar gun during a live fire exercise the night of May 20, 2004. NMCB-22, a battalion based at Ft. Worth, Texas, completed the live fire portion of their military training at Camp Roberts, California. The Seabees attended classes and practiced timed exercises at disassembling and reassembling weaponry for nearly two weeks at Port Hueneme prior to the live fire exercise.

In this photograph from November 16, 2004, Master Chief Constructionman Martin Yingling, Chief Equipment Operator Darion Williams, and Steelworker 3rd Class Justin Sasser stand at attention, wearing Purple Hearts presented to them by Adm. Michael Loose for wounds sustained by indirect fire while working in Camp Fallujah. Yingling, Williams, and Sasser were part of a detachment assigned to Naval Mobile Construction Battalion 4, which was deployed in support of Operation Iraqi Freedom.

Seabees assigned to Naval Mobile Construction Battalion 5 and Thailand Army soldiers work together to build a community center in Ban Poon Suk, Thailand, in this photograph taken on May 10, 2004. Six construction sites were built for the Thai community as part of Cobra Gold 2004. Cobra Gold 2004 was a U.S.-Thai military exercise designed to ensure regional peace through the U.S. Pacific Command's strategy of cooperative engagement. The exercise included land and air, combined naval, amphibious, and special operations as well as U.S.-Thai medical civil affairs projects throughout the kingdom.

A Seabee assigned to the 24th Marine Expeditionary Unit (MEU), pictured here on November 13, 2004, saws off an obstruction in order to apply a seal to the leaks on a broken pipe beside a bridge in Lutafiyah, Iraq. The Seabees and Marines of the 24th MEU rebuilt a bridge after anti-Iraqi militants caused severe damage in repeated attacks. The 24th MEU conducted security and stability operations in the northern Babil province of Iraq.

Pictured June 20, 2005, Seabees assigned to Naval Mobile Construction Battalion 5 pass concrete in buckets in support of a new public works building at Navy Support Activity Bahrain. NMCB-5 was deployed to support maritime security operations (MSO). MSO sets the condition for security and stability in the maritime environment and complemented the counterterrorism and security efforts of regional nations.

Pictured January 10, 2004, members of Naval Mobile Construction Battalion 4 work to level concrete in the foundation of a Habitat for Humanity home located in Oxnard, California.

Steelworker 2nd Class Andrew Blakely, assigned to Naval Mobile Construction Battalion 40, packs his gear on August 31, 2005, in preparation for deploying to Gulfport, Mississippi, in the aftermath of the devastation caused by Hurricane Katrina. NMCB-40 deployed to support Hurricane Katrina recovery in the Gulf Coast as part of Joint Task Force Katrina in support of the Federal Emergency Management Agency (FEMA).

Naval Mobile Construction Battalion 40 load boxes of meals ready to eat (MREs) in preparation for deploying to Gulfport, Mississippi, on August 31, 2005. The Navy's involvement in the humanitarian assistance operations was led by FEMA in conjunction with the Department of Defense.

On September 6, 2005, Seabees attached to Naval Mobile Construction Battalion 40 remove debris from the flightline area aboard Naval Air Station New Orleans after Hurricane Katrina damaged parts of the base.

The same day, members of Naval Mobile Construction Battalion 40 operate a crane to unload equipment and supplies on board Naval Air Station New Orleans.

Seabees observe firsthand the devastation from Hurricane Katrina as they travel along Route 90 in Gulfport, Mississippi, on September 7, 2005.

On September 10, 2005, Navy Seabees assigned to Naval Mobile Construction Battalion 40 transport a truck loaded with a 750-kilowatt generator through the streets of New Orleans to Naval Support Activity East Bank New Orleans.

Pictured September 13, 2005, civilian contractors and Seabees assigned to Naval Mobile Construction Battalion 40 construct a tent city to facilitate 7,500 military and FEMA personnel in New Orleans.

Naval Mobile Construction Battalion 4 removes debris at Our Lady of Lourdes Catholic School's gymnasium in Slidell, Louisiana, during a disaster recovery operation in support of Hurricane Katrina relief efforts, pictured September 11, 2005.

The Seabee statue, designed and created by sculptor Frank Nagy (CM1c), shows two figures representing a Seabee and a young woman at the poignant moment of parting, photographed here on May 10, 1944. The decorative sculptural design was erected in front of Theater A (now the Needham Theater) in memory of the construction battalions passing through Camp Rousseau. The art towers 15 feet above the foundation, inspiring the thousands of enlisted personnel passing near it each day.